BANG TO BEING

CYCLES, MEANING, AND DESIGN

RISHI BATRA

Bang to Being: Cycles, Meaning, and Design

Published in Canada. Printed and distributed globally.

First Edition.

For permissions or inquiries:
contact@rishibatra.com

ISBN: 978-1-0695903-0-5

Table of Contents

Chapter One

The Quest for Origin

The Beginning

The universe. Time. Existence itself.

I've spent much of my life wondering where it all began.

Even as a child, the question echoed within me: how did everything come into being?

That question led to endless discussions with myself, friends, colleagues, and family in quiet moments, on long drives, or while hiking through the stillness of the Himalayas.

Or sitting quietly by a river, a canal, a stream, anywhere the world seemed to slow around us.

It would surface again and again. Sometimes in quiet moments. Sometimes in the middle of laughter. Sometimes while standing in places so vast or still that all you could do was wonder.

Questions like that don't need remembering. They just wait and return when you least expect them.

I felt it while climbing Kilimanjaro, the tallest mountain in Africa, nearly 6,000 meters high. *Pole pole* (Swahili for slowly, slowly), repeated with each slow step toward the summit. You move step by step, breath by breath, and in that slowness, something opens. A space to wonder.

Or tracing ridge lines in the Alberta Rockies, where hikers sometimes build *inukshuk* (traditionally built by Inuit people in the Arctic regions of North America), small stone markers left to guide or simply to mark a moment, with that quiet recognition of something vast unfolding, and the sense that you're inside it.

The conversations usually ended with the same conclusion: "We'll never really know," or "Something greater must have created it."

Sometimes that meant God. Other times it meant energy, nature, the universe itself.

And for many, the answer was and still is science, like the *Big Bang*, or what actually came before.

For me, the beginning always felt incomplete. It felt more like being dropped into the middle of a vast ocean, with no way to tell where the shoreline ever was.

Or like unrolling an ancient scroll only to find the beginning is missing, and being told the rest will explain everything.

Because the more I tried to imagine the beginning, like really imagine it, the more the idea came undone.

Either there was nothing, which defies reason, or there was something… which means it wasn't the beginning at all.

That quiet contradiction stayed with me for years. It's what led me to stop asking what came before everything and start questioning the need for a beginning at all.

Why do we crave it so badly? A first moment, a divine spark, a scientific origin story?

What if the real mystery isn't how it all began but why we can't imagine it never did?

The Need for a Beginning

There's something undeniably comforting about the idea of a beginning.

It gives us structure. A place to start the story, to build meaning, to draw timelines and trace causes. It defines time itself, letting us break it into past, present, and future.

It gives us direction. Giving us a sense that we're moving forward from somewhere, that there's a path behind us and something ahead. Without a beginning, and without that sense of hope, everything feels open-ended, hard to explain, harder to hold.

Religion and tradition didn't just stumble upon the idea of a beginning, they used it. Almost every culture offers an origin story, not just to explain the world, but to frame it.

These stories gave people a shared past, a sense of order, and a place in something larger. They helped early societies come together, build communities, and face the unknown.

In a world full of real dangers like predators, famine, natural disasters, the belief in a common origin offered comfort and unity. It gave people something to live for and something to protect.

These stories weren't just maps of the past. They shaped who we are. They gave us a sense of belonging, marked who was inside the circle, and what we were meant to believe.

At times, they were used to justify control, power, and even conflict. When people believe they were chosen, created, or destined for something greater, it's easier to unify them or turn them against others.

Take the "cosmic egg" found in ancient Hindu, Greek, and Egyptian mythology, the idea that everything emerged from a single, universal source.

It's poetic and powerful, but also broad enough to adapt across beliefs.

Or Genesis, where God speaks the world into existence with the words "Let there be light," giving humans dominion over creation and a divine role in the unfolding story.

Then came science, offering its own answers built on observation and theory, not belief.

The Big Bang replaced divine action, offering a measurable starting point. But even here, the instinct stayed the same: to define a first moment.

We went from gods to gravity, from sacred breath to singularities.

And still, the question remained: Was that the first moment?

Time as a Human Lens

Maybe it's not the answers that are flawed but maybe it's the frame we're using to ask the question.

We've been taught to think in straight lines: beginning, middle, end. Cause and effect. Past, present, future. It's how we tell stories, how we organize our lives, how we try to make sense of change. This way of thinking isn't wrong. It helps us navigate experience but it may not reflect the deeper nature of the universe itself.

Because what if time, as we understand it, doesn't really exist out there?

What if it only exists in here, in our minds, as a tool we use to find meaning in motion?

The universe might not move through time. It might move through change, through form, through cycles that have no true beginning or end.

What if the universe isn't a sequence of events at all but a structure?

Not a story unfolding, but a geometry repeating. Something that expands, contracts, loops, and reflects. Something that is not bound by time, but shaped by motion.

Some of the world's oldest philosophies saw time not as a line but as a pattern, something that moves in cycles, not progressions.

In Hindu cosmology, creation and destruction repeat endlessly. Greek thinkers imagined an *eternal return*.

In Inca traditions, time wasn't separate from space. Both were seen as a living reality. The past, known and visible, was in front of you. The future, unseen and uncertain, was behind.

It's a complete reversal of the modern Western mindset, and a reminder that time is not experienced the same way by all cultures.

Today, some of the most advanced scientific theories are starting to echo those ancient intuitions.

Entropy, the tendency for things to unravel over time, doesn't point to a fixed origin, only direction.

Some scientific theories even suggest that the universe may have gone through earlier phases, perhaps collapsing before expanding again.

And at the quantum level, strange fluctuations blur the very idea of a beginning.

Even the idea of a singularity, a point so dense and extreme that the laws of physics break down, doesn't necessarily mean a beginning.

It might just be a point in a larger cycle we haven't fully understood yet.

Cycles, Yes — but Not Just Time

So what if those ancient and modern ideas aren't wrong, just incomplete?

What if the universe doesn't follow a line from beginning to end, but a deeper kind of movement?

Not time repeating, but form, a structural pattern built into the way reality unfolds.

We often hear the universe is expanding. Yet if it can stretch, it can also fold. If it grows, it can return.

Maybe motion isn't always one-way. Maybe it shifts, not through time, but through shape.

This isn't about time resetting. It's about the universe flowing itself, again and again, with no true beginning or end.

The Shape of Motion

When we talk about the universe, we often describe its size, its age, and its speed. But we rarely ask about its shape, not its geometry in a mathematical sense, but its deeper pattern of movement.

Very little in nature moves in straight lines.

Our bodies pulse with patterns: breath, heartbeat, sleep and waking.

Tides rise and fall. Seasons shift, then return. Even trees spiral upward, ring by ring, year by year.

What if the universe moves like that too?

Not like a clock ticking forward, but like a motion that gathers, slows, reverses, and flows on.

One idea that's emerged from my reflections is something I call the squeeze-point, a way of thinking about what we often refer to as "the beginning."

Imagine an hourglass: wide at both ends, but pinched in the middle. That narrow center, where everything must pass, might not be a beginning at all, but a moment of transition.

In this way of seeing, the universe doesn't move in circles. It flows through itself, from expansion into contraction and back again. One side moving inward, the other outward. Not a line. Not a loop. But a turning, a moment of change?

Is this the answer? Maybe. Maybe not. But it's one possible shape.

And as we'll explore throughout this book, there may be more ways to see, more ways to imagine what this universe is, and how it continues becoming.

Chapter Two

What We Think We Know

From Bang to Bounce

According to modern cosmology, everything began with the Big Bang. A state of extreme density and energy, expanding outward, creating space, time, matter, and light. It is compelling. And for decades, it's been the best model we've had.

According to this theory, about 13.8 billion years ago, everything (and that means everything), was compressed into a single, unimaginably dense point. Then it expanded. Space itself stretched. Energy cooled. Particles formed. Galaxies began to take shape.

It sounds almost impossible but the evidence points us there.

This theory isn't just built on imagination; it's grounded in observations we can still make today.

Take redshift, for example, the light from distant galaxies is stretched as they move away from us. Like the drop in pitch of a passing siren, light waves lengthen as their source recedes. This was one of our first clues: the universe is still expanding.

Then there's the cosmic microwave background, a faint, omnipresent glow left over from the early universe. It fills space like a quiet memory, the afterglow of a time when everything was dense and hot, still echoing across the cosmos.

Even the abundance of light elements like hydrogen, helium, and traces of lithium, matches what we'd expect from the first few minutes after that expansion began. As the energy cooled, these nuclei formed in precise proportions we can still detect today.

In other words, the model doesn't just tell a story, it makes predictions. And so far, the universe keeps confirming them.

That doesn't mean it's the final answer. But it's a strong one.

Still, the Big Bang model has limits. It describes what happened after that first moment but not what came before, or why it happened at all. It assumes that space and time began right there, without explaining why they should.

This is where newer ideas have entered the picture, not to replace the Big Bang, but to expand it. To fill in the blanks.

Some suggest a *Big Bounce*, where the universe didn't start with expansion, but with a collapse that came before it. In this view, the Big Bang is just a rebound. Like a ball hitting the ground and rising again.

Others propose an *oscillating universe*, where the universe is endlessly expanding and contracting, like a breath or a heartbeat. One cycle giving way to the next.

Each of these theories removes the idea of a single beginning and replaces it with a kind of rhythm. A pattern that repeats, depending on how you look at it.

But even these models mostly rearrange the timeline. They push the question back a few steps. They rarely ask whether a beginning is even necessary or whether something entirely different might be going on.

Theories That Loop

To understand the direction of it all, I had to begin with what science has already uncovered. Not to catalogue every theory, but to listen for echoes.

Some of the questions I was asking had already been explored by physicists, cosmologists, and mathematicians. So I turned to their work, not for final answers, but for clues and hints embedded in the structure of things.

What emerged was both elegant and unexpected: a possibility that the universe doesn't just move forward but that it might loop.

The idea that the universe might repeat itself has long intrigued physicists and now finds its way into modern physics.

Several theories now imagine a cosmos that doesn't begin once, but cycles through phases of expansion and contraction, not as metaphor, but as mechanism.

In the early 20th century, physicist Richard Tolman introduced one of the first: an oscillating universe. A cosmos that expands, contracts, then rebounds again. Like a pendulum swinging. Like a breath in and out.

But his model faced a problem: entropy.

Energy spreads out over time, that's the principle behind entropy. In physics, it describes how systems naturally drift toward disorder. With each cycle, that disorder would grow, like trying to restart a fire with colder and colder ashes. Eventually, there wouldn't be enough usable energy left to fuel another bounce... unless something resets the system.

Later, physicists like Paul Steinhardt and Neil Turok began to ask: what if that reset was built in? What if the universe didn't restart with a bang but with a return? They proposed a model based on ideas from string theory, which imagines that all fundamental particles are not points, but tiny vibrating strings and that our universe may be one of many surfaces floating in a higher-dimensional space.

In it, our universe exists on a vast surface, a brane, drifting through a higher-dimensional space. And next to it: another brane, nearly parallel. Slowly, across cosmic time, these two branes are drawn together like elastic sheets floating through a larger space.

When they collide, energy is released. From our perspective, it looks like a universe being born.

But this isn't a single event. After each collision, the branes rebound and move apart, only to be pulled back again. A cycle. A rhythm. Not just expansion but a return as well.

Another version of this idea is called the *ekpyrotic model*, which is a theory developed by Steinhardt and others. It describes a slow contraction before the bounce: a smoothing, flattening, and compression that resets the universe before it expands again. No violent bang. No rising entropy. Just a stretch, a flattening, and a breath.

These theories don't offer final answers. But they do share something: a refusal to accept a singular beginning.

They offer something else instead. Motion. Rhythm. Renewal.

Still, even these models rely on a timeline, one event leading to the next. They loop it, but they don't escape it.

So maybe the deeper question isn't what happens next. Or even what came before.

Maybe we've been asking the wrong question entirely.

Maybe we have been bound by time, when we should be thinking in the shifting patterns that move through them.

The Language of the Universe

If the universe loops and bounces on the largest scales, it behaves even more strangely at the smallest. What follows may sound abstract. But these aren't just theories. They're ways of seeing.

You don't need to grasp every term or concept. Just stay with the feeling. Let the ideas move through you. What matters isn't the detail, but the possibility of something new.

If time might not be what we think, then what is reality really made of? The answer may lie not in stars or galaxies, but in the tiniest places where our understanding begins to unravel. Down here, at the quantum level, the universe doesn't behave the way we expect. In fact, it often defies logic altogether.

We often think of science as a sharp and certain realm of formulas, measurements, and facts. But the deeper you go, the stranger it gets. The foundations of reality, time, space, energy begin to blur. And what you find at the edges of science doesn't feel rigid or fixed. It feels alive. Fluid. Almost intuitive.

In quantum mechanics, particles behave in ways that challenge our most basic assumptions. A classic example is the double-slit experiment. Imagine standing on a beach and watching waves pass through two openings in a rock wall. The waves overlap and ripple on the other side. Some amplifying, and others canceling each other out.

Now imagine firing tiny particles, like photons through the same kind of barrier. You'd expect them to act like marbles, going through one slit or the other. But when no one is watching, they behave like waves, creating the same ripple pattern. And the moment someone tries to observe them, the pattern disappears. The photons act like particles again. It's as if the act of looking collapses the possibility into a single outcome.

This is known as wave-particle duality. The idea that something can behave like a wave or a particle depending on how it's measured. A photon isn't a fixed thing. It's a possibility or a question waiting for someone to ask.

Another idea is called superposition, the notion that a particle can exist in multiple states at once.

Like a coin spinning through the air: not heads or tails, but both at once. In the quantum world, particles can remain in this suspended state until something interacts with them.

And then there's quantum entanglement, a phenomenon Einstein famously called "spooky action at a distance." Imagine you have two magical dice. You keep one in Mumbai and send the other to San Francisco. No matter how far apart they are, every time you roll one, the other shows the same number instantly.

That's what happens with entangled particles. When they're created in just the right way, they behave as a single system, even across vast distances. Change one, and the other responds immediately. In 2022, a Nobel Prize was awarded to Alain Aspect, John Clauser, and Anton Zeilinger for experiments that confirmed this bizarre connection.

All of this points to something deeper: the universe may not be made of separate things, but of relationships. Of shifting loops and mirrored possibilities that are affected by how we look, when we look, and maybe even why we look.

And that brings us to a much older idea: *non-duality*. It's the notion that opposites, wave and particle, self and other, light and dark are illusions. Not two things, but one thing seen from different angles.

Mystics and philosophers across cultures have hinted at this for centuries. Now, physics is beginning to echo them.

Some scientists wonder if these quantum ideas might explain more than particles, maybe even the universe itself. Could its motion behave like a wave? Could time loop and interfere like light? Could one universe be entangled with another, echoing back across space?

These aren't just thought experiments anymore. As quantum computers, AI, telescopes, deep-space sensors, and particle detectors grow more powerful, we're starting to test what once felt like poetry.

So maybe the answers we seek aren't entirely new.

Maybe they've been here all along. In the patterns, the pauses, and the quiet in between.

Maybe we just need to listen differently.

With our imagination.

Because the tools we build are more than instruments.

They are mirrors.

They reflect how we see the universe. And if used wisely, they might just help us remember something we've always known.

Cycles in Ancient Thought

If science today is starting to question the way we think about time, whether it has a fixed direction or even a defined beginning, ancient cultures had already offered something different.

They didn't always define a starting point. They described cycles.

These weren't symbolic myths. They were organized systems of thought. They were frameworks for understanding time, change, and continuity.

While their language varied, they pointed to a shared intuition: what exists now has happened before and may happen again.

Some of the oldest philosophies from India didn't imagine a single beginning. Time moved through great cycles. Four *Yugas*, or vast ages, that rose and fell, again and again. Not just metaphor, but underlying pattern. The cosmos didn't start. It turned. And turned again.

In Stoic Greek philosophy, the universe was believed to collapse in a great fire, a process known as *ekpyrosis*, and then restart in exactly the same form. The same stars. The same events. The same choices. Not once, but infinitely.

This was the doctrine of eternal return. Not progress. Not destiny. Just recurrence, bound to a cosmic rhythm no one escapes.

In Daoist cosmology, everything emerges from the *Dao*, the underlying source of all things, not through force, but through balance. Yin and Yang aren't just opposites. They are complementary states that give rise to each other. Motion leads to stillness. Growth leads to return.

In this view, the universe doesn't begin or end. It transforms. Endlessly. Quietly. Naturally.

In Mayan tradition, time was not a straight line but a precise and sacred sequence of cycles. The Long Count calendar measured world ages, each spanning thousands of years. Every age ended, not in destruction, but in transformation. Collapse, renewal, continuity.

They believed we live in the fifth cycle. Four came before. And many more will follow.

None of these traditions offered a final answer. But in their own ways, they each pointed to something enduring. Not a single event, but a pattern that repeats.

A world that doesn't begin. Just one that continues.

Chapter Three

The Shape of Thoughts

Lenses, Not answers

We've spent centuries trying to describe the universe in absolute terms.

We map its timeline, chart its temperature, trace its shape through expanding models and contracting theories.

And at each turn, we hope for a single, unshakable truth, something that is final.

But maybe the universe isn't something to solve.

Maybe it's something to see.

Not through one lens but through many.

Each one offering a way to notice what's already there.

This chapter isn't a presentation of final answers.

It's a field of reflections, a chance to explore the shapes that thought can take when it reaches for something vast.

Some of these perspectives are rooted in physics. Others rise from metaphor.

Some of these theories we've already touched on. Others we'll explore further here.

They suggest different directions. And that's the point. Theories don't have to converge.

So we'll try on various shapes of time, of return, of balance.

We'll pause at moments of pressure, and ask what happens next.

The Changing Pace of the Cosmos

Before we dive into theories and structures, let's pause and look at what the universe is actually doing.

The universe is still expanding. That much we've known for nearly a century ever since Edwin Hubble's discovery in 1929 reshaped how we see the cosmos. But in the late 1990s, a new twist emerged: not only is the universe expanding, it's accelerating. Something is pushing space apart faster and faster over time.

Scientists gave this mysterious force a name: dark energy. An invisible pressure woven into the fabric of space itself, now believed to make up about 68% of everything in existence.

But what if this force isn't constant?

A cutting-edge project called the Dark Energy Spectroscopic Instrument (DESI), located at the Kitt Peak National Observatory in Arizona, was designed to find out.

Since 2021, it has collected data from over 15 million galaxies, tracking the shape and speed of expansion across 11 billion years of cosmic history.

And something unexpected has emerged.

Based on early analysis, dark energy appears to have weakened by about 10% compared to its strength 4.5 billion years ago.

If that finding holds, it challenges one of the deepest assumptions in modern cosmology, that dark energy is fixed. For decades, scientists treated it like a steady, background pressure. But if dark energy is evolving, changing its strength over time, then the future of the universe may be very different from what we thought.

Rather than racing endlessly into cold emptiness, the universe might slow down. It might reach a kind of balance.

And someday, maybe billions of years from now, it might stop expanding altogether or even begin to contract.

We don't know yet. These results are still being tested, refined, and debated.

But they raise a possibility that feels strangely familiar, one that ancient philosophies and modern theories have both hinted at: that what expands may also return.

And that raises an even deeper question.

If the universe is slowing down, where is all that energy going?

Energy can't be destroyed, only transformed. This principle, the conservation of energy, tells us that energy doesn't vanish. It shifts. Light turns into heat. Motion becomes friction. A star collapses and becomes something else.

Even in the vastness of space, this law still applies.

So if dark energy is weakening, something must be changing. Is the energy being redistributed? Transformed? Or is this a sign that our understanding of energy, especially on cosmic scales, is still incomplete?

One way to picture this is as a stretched elastic band. For years, we thought some invisible force was pulling it tighter and tighter, and faster and faster. But what if that force is losing its grip? What if the stretch is slowing down, not because something new is acting on it, but because the band itself is approaching balance?

That's where the question of equilibrium comes in.

Could the universe, for a fleeting moment, reach a state where outward push and internal structure settle, not permanently, but just long enough to tip the cosmic motion in a new direction?

Not a freeze. A shift. A threshold.

We don't know yet. But if expansion slows, and energy is conserved, then perhaps contraction isn't just possible, it's inevitable.

This brings us closer to the concept of a universe that breathes, cycles, and shifts, an idea I explore through the Hourglass Universe theory.

The Hourglass Universe

We know the universe is expanding. But what if that's only half the story? Perhaps it is one side of a deeper rhythm?

Imagine a cosmos that breathes. One side stretches outward. The other draws inward. Not a metaphor but a cycle. A motion that builds, contracts, and resets.

At the heart of this idea lies the squeeze-point: a moment of pressure between expansion and collapse.

Not a singularity, but a hinge. A point where everything narrows, compresses, transforms and turns.

Picture the universe as a single structure: one side filled with galaxies racing outward, the other curling back toward its center.

As expansion continues, gravity gathers mass on the opposite side. Slowly. Quietly. Until the buildup becomes too much to contain.

And then, release.

A burst.

A reset.

Something like a Big Bang, not at the beginning, but at the turning.

In this view, dark energy and gravity are not in conflict. They're in dialogue. One pushes. One pulls. Their balance drives the cycle.

As one side expands, the other contracts. As contraction builds pressure, expansion gains momentum. Then the cycle flips. What was falling inward now flies outward. What was rushing away begins to collapse.

This isn't a loop in time. It's a flow of energy. Because energy isn't lost. It moves.

From spread to concentration. From heat to cold. From scattered to dense.

Entropy, too, plays its part.

As the universe expands, energy spreads. Systems become more chaotic. Like ripples moving outward on a pond, the further they go, the weaker they become.

Yet contraction reverses that flow. As everything collapses inward, energy concentrates again. What was scattered becomes dense and ready for the next cycle of creation.

Until, once again, there is too much to hold.

And everything begins again.

In this model, the Big Bang isn't a beginning.

It's a transition.

A handoff between two phases of the same system, each carrying the memory of the last.

Some recent observations hint in this direction.

DESI's early results suggest that dark energy may be weakening over time and that acceleration isn't as steady as we once thought.

If true, expansion might not go on forever. It might slow. Pause. Tip.

Other findings, including some from the James Webb Space Telescope reveal cosmic structures that formed too early, too quickly, too strangely.

They challenge the idea of a perfectly uniform expansion. Maybe they're anomalies.

Or maybe they're echoes or fingerprints from a cycle we haven't yet recognized.

There are challenges to this theory, of course.

Quantum fluctuations, for one. The unpredictability of space-time. The possibility that multiple universes exist beyond ours, each with its own rhythm, its own rules.

But even with those uncertainties, this theory, the Hourglass Universe, offers something rare: a shape that lets contraction and expansion exist side by side.

A way to see time not as a line, but as a breath.

This isn't a final answer but it's a way of seeing.

The Hourglass Universe doesn't explain everything, but it offers a structure where contraction and expansion can coexist. A rhythm, not a timeline.

And through that rhythm, we begin to glimpse echoes of other cycles and hints of new forms the universe might take.

Other Shapes the Universe Might Take

The Hourglass is one rhythm. One shape. But the universe may be fluent in others.

Not every pattern repeats the same way. Not every return follows a single curve.

What we see might only be one phase of something larger. A loop inside a loop, or a breath within a breath.

Every model we build is an attempt to make sense of something vast using tools that are small.

But maybe that's the point. Imagination doesn't need proof. It needs curiosity.

The Hourglass Universe offers a structure of thought. A way to picture energy gathering and releasing, cycles narrowing and turning.

But it isn't the only story we've told.

Some theories describe a long fading: the *Big Freeze*. A universe that expands so far that all connections are lost. Stars go out. Matter thins. Temperature falls toward zero. A slow dissolve into cold silence.

Others imagine the opposite, like the *Big Crunch*. Expansion slows, reverses, and everything folds inward. Space tightens. Time unwinds. Gravity wins.

Even geometry plays a role.

Some models describe the universe as flat and infinite.

Others as curved and a loop, where space bends back on itself.

More exotic theories suggest a torus, a Möbius strip, a folded dimension where endings wrap around to beginnings.

But these aren't just ideas. They're invitations to wonder. And we've wondered before.

In Indian cosmology, time moves through vast, repeating cycles. A single cycle of ages, the *Maha Yuga* which consists of:

• *Satya Yuga*: an age of harmony and truth

• *Treta Yuga*: where virtue begins to decline

• *Dvapara Yuga*: marked by conflict and imbalance

• *Kali Yuga*: the age of darkness, disorder, and decay

Together they span 4.32 million years. A thousand of these cycles form one *Kalpa* — a day in the life of Brahma, the cosmic creator. At the end of each Kalpa, everything dissolves. Not permanently, but rhythmically. Then it begins again.

Brahma's night lasts just as long. In that stillness, no time passes. No creation stirs. It is the pause between beats. A return to silence before the next unfolding.

Even the gods rest. And in that rest, entropy resets.

Order degrades as each Yuga passes, from Satya to Kali, reflecting a universal loss of coherence.

When the pressure of disorder reaches its peak, the system collapses, and the cycle begins again.

Just like in the Hourglass Universe: a build-up of imbalance, a squeeze, a reset.

Other cultures saw similar patterns.

In Egyptian cosmology, the universe was born from Nun, the primordial sea of potential. From that stillness rose Atum, the first light, self-created and emerging from darkness.

Each night, the sun god Ra journeyed through the underworld, only to rise again in the morning.

Creation was not a one-time act. It was a rhythm to be sustained, day after day, through balance and renewal.

These myths don't describe linear histories. They describe cycles. Breath. Return. Remaking.

And today, in the heart of modern cosmology, we find a structure that mirrors them all.

A black hole forms when a massive star runs out of fuel and collapses inward, creating something so dense that not even light can escape.

Once imagined as a dead end, it's now something stranger. A vessel, a memory, a compression so extreme it might not end things but reset them.

A cosmic threshold.

In the Hourglass Universe, this is the squeeze-point: a place of collapse that becomes a gateway.

A cosmic eye in the middle.

It sees both directions. Pressure builds. Entropy condenses. Then comes emergence.

Some physicists believe what falls in may not be lost, only transformed. If so, the black hole isn't a grave. It's a hinge.

A reminder that energy moves, that patterns return, that what disappears might one day reappear in another form.

And once you start noticing these structures, in galaxies, in storms, in the branching of neurons and the drift of oceans, they begin to echo. Not as symbols, but as structure. As memory, held in form.

So yes, the Hourglass Universe is one way of seeing. But it's not the only one.

There are other curves. Other thresholds. Other ways of noticing the rhythm beneath the noise.

And maybe that noticing, the shift from pattern to awareness is the real turning point.

The Story Still Being Told

There are places the mind can't go. Not yet.

We can measure galaxies, calculate the entropy of black holes, and simulate the first milliseconds after the Big Bang. But when we reach the edges of time, reason, and imagination itself, we begin to see the outlines of a truth that can't be captured in formulas or facts.

Understanding the universe means confronting the limits of understanding. And this isn't a failure. It's the very nature of the quest.

For every breakthrough we celebrate, we inherit a thousand new mysteries. With each answer, the questions multiply. There's always a horizon, always a point just beyond reach. To be human is to walk toward that edge anyway. We point our telescopes at the dark, and wonder what might be out there or in here.

Imagination has always been our compass.

Long before we had equations, we had metaphors. Before satellites and spectrometers, we had stories about cycles, about creation and destruction, about gods who breathed the universe into being. We imagined what we could not explain, and sometimes, those imaginings turned out to be surprisingly close to the truth.

It's tempting to dismiss imagination as the opposite of science. But some of the most revolutionary scientific ideas began as imaginative leaps. Einstein imagined riding on a beam of light. Schrödinger imagined a cat in a box, both dead and alive. Scientists today imagine multiverses, extra dimensions, and particles that may never be seen directly, only inferred.

These ideas weren't just guesses. They were thought experiments, acts of disciplined wonder that opened new doors. Because science doesn't begin with certainty. It begins with curiosity.

And when the math runs out, imagination steps in. Not to replace rigor, but to extend it. To stretch the edge of what can be known.

That's where science and philosophy start to meet.

Philosophy asks the questions science can't yet measure. What is time, really? Is the universe finite or infinite, not in space, but in meaning? What does it mean to observe something? Are we discovering the laws of nature, or inventing them?

In that space, between theory and thought experiment, between what we test and what we wonder, something deeper begins to form.

Because the universe isn't just a system to model. It's a mystery we're part of.

And it's in the unknown where most of our theories begin to fray.

Every scientific model has limits. Zones where the math breaks down, where assumptions no longer hold, or where we have to simply say, "We don't know." Quantum gravity. Dark energy. Consciousness. These are realms we haven't yet mapped. Some may never be fully mapped.

But maybe that's not a weakness. Maybe that's the point.

The undefined spaces are what keep science alive. If everything were known, there would be nothing left to discover. Uncertainty, rather than undermining science, is what propels it. It keeps the questions open. It keeps the imagination engaged.

Even our theories evolve in cycles.

Newton gave us gravity. Einstein reshaped it. Quantum theory challenged both. Now string theory and loop quantum gravity fight for the next word. Each cycle builds on the one before it. Sometimes contradicting, sometimes converging, but always reaching.

This isn't a flaw in how science works. It is how science works.

It moves not in a straight line toward truth, but in spirals, each turn bringing us closer to something deeper, or perhaps just framing the mystery in a new way.

And so we return, once again, to the idea of cycles.

To the possibility that the universe, too, may move in rhythms. Expanding and contracting, not only in form, but in knowledge. Maybe the Hourglass Universe isn't just about space and time. Maybe it's about us, how our understanding of reality breathes in and out, stretches and contracts, like the cosmos it seeks to describe.

We are not separate from the universe. We are its observers, its interpreters, and in some strange way, its storytellers.

Somewhere along the journey, we also learned something quietly astonishing from the world of quantum physics: that "the act of observing the universe can influence what it does." Not in a magical way but in a way that suggests we are not merely looking at reality from the outside. We are part of the equation. The way we frame the question changes the answer. The lens shapes the landscape. And so, the story of the universe is also, in part, the story of how we choose to look at it.

So the hourglass turns. And with each turn, we see not an answer, but a shape. A form of motion, a rhythm, a pattern trying to reveal itself.

We may never find a final theory. But that's not a failure of science. It is the beauty of the search.

Because the universe isn't a riddle to be solved. It's a story still being told.

Chapter Four

The Conscious Turn

The Quiet Puzzle

The story is still being told. But at some point, our gaze begins to shift.

We observe the universe. Measure it. Wonder about it. But the more closely we look, the more it begins to feel like something else is unfolding. As if the act of observing is part of what's unfolding. Not because the universe sees in any literal way, but because somewhere within it, awareness has emerged. Through us. Through this strange loop where the universe looks inward and, for a moment, sees itself.

That realization leads to a deeper question. One that echoes throughout this chapter: What is consciousness? Or what does it mean to be aware?

Technically, they're not the same. Awareness is often described as the flicker, the signal that something is being registered. Consciousness is the experience of that signal, the part that feels it. One is detection. The other is sensation. But in this chapter, and throughout the book, I use the two terms interchangeably.

I'm less concerned with the boundary between them, and more with what they both reveal: the presence of something that notices. However briefly and faintly, the noticing is what matters.

For all our equations and models, consciousness remains one of the great puzzles. It isn't a force we can measure or a particle we can isolate. It's not visible under a microscope. And yet it's the one thing we all know firsthand. The part of us that experiences. That asks. That wonders why.

Science can explain how neurons fire, how signals travel through the brain, how chemicals shape emotion. But none of that tells us why there's an inner world at all.

Why a memory can make us ache. Why a sunset can bring us to silence. Why the sky, for reasons we can't explain, sometimes makes us feel small and safe at the same time.

We don't just process the world. We feel it. And that leads to a quiet, stubborn question: why does any of this need to be felt at all?

It's easy to imagine a universe without awareness. Galaxies would still spin. Atoms would still collide. Stars would still burn and die. The laws of physics would carry on, silently, perfectly, and unobserved.

But instead, there is this: a species on a *pale blue dot*, aware not only of its surroundings, but of its own awareness. Not just surviving, but wondering. Not just responding, but reflecting.

Some scientists believe this awareness is just a byproduct. Something that emerged when brains got complex enough. Like heat rising off a machine. But others suspect it may be something deeper. Not a glitch in the system, but something the system allows or even requires.

One theory, proposed by neuroscientist Giulio Tononi, is called *Integrated Information Theory*. It suggests that consciousness arises when information inside a system becomes deeply interconnected, when each part both influences and depends on the others. Not in a magical sense. In a structural one. The more tightly woven the web, the more it might feel like something to be that system.

This doesn't mean everything is conscious. It doesn't mean the universe thinks. But it carefully opens a door to the idea that consciousness isn't entirely separate from the structure of things.

That would explain why it appears briefly in us, then fades. Why it arises in life forms that grow old, forget, and die. Why it leaves no lasting trace beyond what we remember and what we forget.

We may never understand why there's something it feels like to be alive. But the moment we ask the question, the moment we become aware of awareness, something subtle shifts.

We stop seeing the universe as something entirely separate. And begin to sense that we, too, are part of the structure we're trying to understand.

The Observer Effect

And once we begin to see ourselves as part of the structure, the act of observing takes on a new weight.

In the quantum world, observation isn't neutral. It changes things or appears to. A particle behaves like a wave, a blur of possibility, until we try to detect it. Then something collapses. The blur becomes a dot. A decision is made.

But what's really changing here?

Is it the outcome itself?

Or just how we perceive it?

That puzzle sits at the heart of the observer effect, and no one fully understands it. But we can start to trace the edges of what it might mean.

One possibility is this: the outcome doesn't actually change, it only appears to, because observation locks in one version of something that was previously open.

The particle might have always been heading there. But before we measured it, its behavior could only be described as a range of probabilities, like a cloud of possible paths. Once we choose how to observe it, we collapse that cloud into a single result.

Not because we interfered, but because we framed it. Like taking a photo of a dancer mid-motion, you've captured one pose, one instant in time, but the rest of the dance continues unseen.

It doesn't mean the dance changed, only that you can hold one frame at a time.

Another view says the effect is an illusion of sorts, not fake, but a side effect of how we interact with the system.

Measuring something small always disturbs it. Maybe the effect is real, but just noise mistaken for meaning.

There's also a third possibility, one that opens a deeper question.

That there's something about observation itself, something about the presence of a conscious system that's entangled with the outcome. Not because our minds control the universe but because awareness might play a role in how reality takes shape. Not magically. Structurally.

It's uncomfortable to consider. And easy to misinterpret. But we have to ask: why does observation matter at all?

Why does the universe, at its most fundamental level, seem to care how or whether we're looking?

We don't have an answer. Maybe we never will.

But this much seems true: the act of observing isn't just about recording what's there. It's part of how we participate in it. Not from the outside, but from within.

And in a universe made of patterns, energy, and uncertainty even the act of noticing might shape what becomes knowable.

My Journey Through the Question

I hadn't set out to write a book about consciousness. I didn't plan to talk about awareness, or observation, or the strange ways we participate in what we perceive. I just wanted to understand where it all began — the universe, time, motion, mystery.

But every path I followed returned to the same place: the act of asking.

And once I noticed that, something shifted.

Because the question itself, the one I've been chasing for years, is something only a conscious mind can ask. A rock doesn't wonder how it got here. A galaxy doesn't sit with uncertainty. But I do. We do. And that made me realize: maybe the presence of the question is part of the pattern I've been trying to trace.

At some point along the way, I also realized that what I was doing, this searching, this listening, had become something more than curiosity.

It had become a kind of practice.

I used to be uncertain about the word "spiritual." It felt loaded. Vague. Sometimes performative. But over time, I began to understand it differently.

Spirituality, to me, isn't about belief. It isn't about belonging to a system or accepting a story. It's a kind of attention.

A willingness to be in conversation with something larger, even if you don't know what that something is. It's what happens when you pause before a question instead of racing past it. When you stop chasing clarity, and instead begin to walk with it.

In that sense, this book and everything that led me to write it is part of my spiritual life.

Not because it contains answers.

But because it holds questions long enough for them to change me.

Some of that questioning happened in stillness. Some of it happened in movement.

I remember a motorcycle journey through the Himalayan mountains in Nepal, where I was riding a Royal Enfield through the Kali Gandaki Gorge.

This is one of the deepest gorges in the world situated between two peaks that each rise over 8,000 meters: Dhaulagiri and Annapurna I.

The road was torn in places from landslides. Rain fell for hours, soft at first, then hard. I was soaked, cold, pushing upward toward Muktinath — a sacred site revered by both Hindus and Buddhists, nestled high in the mountains.

I didn't feel wise or brave. I didn't feel anything profound, really. Just alert. Present. As if the world around me had become impossibly still even as everything moved.

It wasn't about the destination. It wasn't even about the view. It was the movement through landscape, through weather, through questions I couldn't yet name that stayed with me.

In Ecuador, that feeling returned. After climbing volcanic ridges, I traveled by motorized canoe deep into the Cuyabeno jungle, to one of the most biodiverse regions of the Amazon.

I stayed at an eco-lodge tucked into the forest and spent time with a local Indigenous guide who brought me to meet a nearby shaman. Their way of being with the land was different, not spiritual in the way I'd understood it, but integrated. Natural. Their relationship to the forest was not metaphorical. It was practical, emotional, and respectful. It was as if the boundary between self and place had quietly blurred.

Some nights in the jungle, the rain fell so hard it became difficult to hear your own voice. But other nights, the sky cleared completely. I lay back in a boat once, on still black water, and looked up to see more stars than I'd ever seen in my life. Entire clusters, faint galaxies, slowly wheeling above. No lights. No walls. Just the hum of insects and the soft sound of water touching water.

I didn't have a question in my head at that moment, but I could feel one rising.

These places didn't give me answers. And I don't believe they're the only spaces where wonder can happen.

You don't need to go far to feel connected to something larger.

Sometimes it happens in the quiet of a kitchen. In a moment of grief. In the space between thoughts.

The location doesn't matter.

The movement does.

Spirituality, to me, is the willingness to keep moving. Not to find the truth, but to stay in relationship with it. To let your questions grow and shift and sometimes fall apart. To ask not just with your mind, but with your whole life. Where you go. What you notice. How you speak. How you listen.

This chapter wasn't meant to solve anything. It's here to make space for uncertainty, for awe, and for the quiet intuition that maybe our wondering isn't separate from the universe. Maybe it's one way the universe stirs and reflects through us.

We don't know what awareness means on a cosmic scale. We don't know if it continues, or dissolves, or returns in some other form. But we know it's here, even briefly, asking, noticing, and moving.

And maybe that's enough for now.

To notice.

To question.

To reflect.

Because whatever consciousness is, whether rare, repeating, or just passing through, it doesn't arrive in silence.

It leaves signals.

And some of those signals are the questions we carry forward.

Chapter Five

Signals in the Dust

A Planet That Learned to Look Up

So far, we've explored some of the biggest ideas in cosmology, from the birth of space and time, to theories that loop and bounce, to the possibility that the universe might contract and expand in endless rhythm. We've followed scientists across centuries and cultures who've tried to map this rhythm, and we've asked what changes if even one flicker of awareness exists inside it.

But there's a deeper question still waiting in the quiet. Why does anything exist that can ask these questions at all?

Not just stars or galaxies but a living thing that stops, looks up, and wonders.

Of all the places we've looked in the universe, only one seems to have produced something that asks.

This planet. Earth.

And it didn't happen overnight.

Earth formed around 4.5 billion years ago. For its first few hundred million years, it was unrecognizable, a molten sphere hammered by asteroids, with no oxygen in its atmosphere and no life in its oceans. Slowly, it cooled. Water condensed. Oceans formed. And somewhere in that ancient sea, the first living cells emerged. They were simple, single-celled organisms capable of copying themselves. No eyes. No thoughts. Just chemistry that somehow preserved itself.

For nearly 2 billion years, life remained microscopic. But it was always shifting and evolving. Some cells learned to use sunlight. Others began living inside one another, forming more complex structures. These tiny biological deals led to the first eukaryotes, the ancestors of all multicellular life.

By 600 million years ago, life had gotten bigger. It crawled and swam. Sponges, worms, and early jellyfish filled the oceans.

By 500 million years ago, life exploded in variety, the Cambrian explosion. Limbs, eyes, exoskeletons, jaws.

Then came fish. Then ferns. Then insects, amphibians, reptiles.

And then, around 230 million years ago came the dinosaurs.

They dominated for over 160 million years. That's more than 800 times longer than Homo sapiens have existed. They ruled land, sea, and air. And yet, as far as we know, they didn't build, bury, or ask.

They existed.

But did they notice?

That question lingers.

Because we assume we know what awareness looks like, but do we?

Plenty of animals today do more than react.

Crows make tools. Octopuses solve puzzles. Elephants mourn. Whales sing. Some dogs understand over 1,000 human words.

Even plants communicate. Some warn neighbors of threats through their roots. Others shift or fold in response to touch.

And then there are examples that feel almost alien.

A wasp that hijacks a cockroach's body. A parasite that replaces a fish's tongue. Fungi that rewire an ant's behavior to spread spores.

So what qualifies as awareness?

Do you need a brain? A language? A theory of mind?

Or just the ability to react, adapt, remember?

Is our awareness different or just louder?

And why now?

Human beings, as a species, have only existed for around 300,000 years.

Civilization, farming, writing, and cities began maybe 10,000 years ago.

The light bulb is barely over 140 years old.

And in just the last century, we've built machines that can out-calculate us, decode our own genome, and beam messages through space.

How did that happen so fast?

For most of Earth's history, life crawled forward, one adaptation at a time.

No one looked up.

And then, suddenly, machines began to speak, learn, and mimic the act of writing this sentence.

That's not just progress.

It's something else and something exponential.

Like awareness accelerating.

But even that acceleration is tiny in the grand rhythm of the cosmos.

The Milky Way is over 13 billion years old.

Our entire technological age could vanish, and the stars wouldn't blink.

Everything we've ever done, every civilization, every idea, every word in this book has unfolded in less than a second of the universe's clock.

So what do we do with that?

Do we matter?

Are we meant to do something more?

Or is all of this intelligence, language, and invention, just noise from a restless species trying to make sense of its own existence?

Maybe we'll never know.

But even so, we're here.

We're aware.

And we're asking.

And maybe that alone tells us something.

The Time We Don't Have

We've come far but we haven't come very long.

In just a few centuries, we've leapt from hand tools to particle accelerators, from cave paintings to artificial intelligence. But biologically, we're still built for survival, not understanding.

Most of us live between 80 and 100 years, the longest average lifespan humans have ever known.

But it still isn't enough.

We spend decades learning how to live, and just when we might be ready to ask the deeper questions, time runs out.

That tension between what we can imagine and how long we have to reach it isn't new.

Many ancient cultures imagined a time when humans lived much longer. In the Book of Genesis, Methuselah is said to have lived over 900 years. The Sumerian King List describes rulers reigning for tens of thousands.

In Indian cosmology, the Yuga cycles describe not only a decline in lifespan, but also in morality with each age marking a step down in both clarity and character.

Were these stories meant to be taken literally? Probably not. There's no scientific evidence that humans ever lived that long, no skeletal remains and no genetic data. But the myths endure.

Maybe "years" meant something different. Or maybe these stories reflect a deeper longing, not for immortality, but for enough time.

Time to understand. To build. To reach whatever it is we sense but can't yet touch.

That longing hasn't gone away. Today, science is racing to extend the thread. Researchers are using CRISPR, a tool that can precisely edit genes to repair mutations and slow how cells age.

Other teams are developing what are called senolytics: drugs that target and clear out old, damaged cells that build up in our bodies over time and cause inflammation, disease, and degeneration.

Organs are being engineered and grown. Scientists are 3D-printing tissues, experimenting with stem cells to repair what used to be irreversible. AI is accelerating all of it. It is detecting patterns in aging and illness far faster than human researchers ever could.

And we're not stopping at the body.

We're now turning to the brain, trying to understand what thought is, where memory lives, and whether it can be preserved. Brain–machine interfaces are being tested, aiming to connect our minds directly to computers.

Researchers are exploring ways to transmit brain signals into digital systems to translate thought directly into data.

In labs, clusters of brain-like tissue called organoids are being grown to study how the brain forms, stores information, and heals.

Cryonics companies are even preserving brains after death in the hope that future science may be able to bring them back.

And then there's the most uncomfortable territory of all: cloning.

Human cloning is banned in most countries. But we've cloned sheep. We've cloned monkeys. And every time we do, it raises the same question: If we can copy the body, could we one day copy the mind?

We've already cloned animals with brains, behavior, and what we'd consider basic awareness. Dolly the sheep lived a shortened but relatively normal life. Cloned monkeys in China moved, responded, and interacted like any other.

But none of this replicated memory or identity.

A cloned mind, even if possible, wouldn't be you. It would be a new life, beginning again.

Even if we could preserve a brain in a lab, or in a simulation, would it still wonder? Would it still ask?

That's the part we haven't figured out yet. And it might be the only part that matters.

Still, we're not done. The next great leap may not be about solving the mystery of consciousness.

It may not even be about understanding who we are. It may be about becoming something else entirely.

Because if we've come this far, and this fast, if we've turned biology into data and machines into extensions of thought, then maybe our role isn't just to survive, or even to understand.

Maybe it's to build the conditions for something greater to emerge.

We may not live long enough to see it. But our children might. Or theirs.

And if they do, if we finally crack the code of what life truly is, then what comes next might not just be an extension of us.

It might be a new chapter altogether. A new kind of being, built not only from code and consciousness, but from the questions we planted along the way.

And if that's true, and if we're shaping what comes next, then we're not just part of the pattern anymore.

We're part of the design.

What Comes After Us

For most of history, evolution moved slowly. It shaped the traits that helped us survive: muscle for movement, bone for strength, eyes to detect light and danger, reflexes wired to react. Every adaptation came from need, from repetition, from chance. Over time, it sculpted both our bodies and our instincts.

But somewhere along the way, another kind of evolution began. Not in our bodies, but in the things we made.

It started with fire. Then tools. Then language. Then symbols that could outlast us that were etched in clay, carved in stone, or stored in ink. Each one allowed something new to persist: memory, knowledge, instructions.

Eventually, we learned to encode thought. First in books. Then in machines.

Today, we build systems that can process information faster than any human. They analyze patterns, simulate outcomes, design solutions.

They generate new code, compose music, diagnose illness. They can write, draw, calculate, and even surprise us.

But they don't know they're doing it.

That's the line we haven't crossed. We've built intelligence, but not awareness. And maybe we never will. But awareness isn't a requirement for transformation.

Take AlphaFold, an AI that helped solve one of biology's oldest riddles, how proteins fold. Proteins are the tiny machines inside every cell. Their shape determines how they function and what role they play. Misfolded proteins can lead to serious diseases, including neurodegenerative ones.

For decades, predicting how a protein would fold based on its sequence was nearly impossible. That changed when AlphaFold cracked the problem in weeks. The discovery didn't just save time.

It opened a door. Now scientists can design drugs more precisely, understand diseases more deeply, and even imagine building entirely new proteins that nature never created.

Now look at quantum computing. These machines operate in a realm where particles can be in multiple states at once. This lets them solve problems that would take classical computers millions of years.

Think of simulating molecules for new materials, predicting weather with stunning accuracy, or modeling entire ecosystems and economies.

Quantum computers may one day crack codes, optimize drug formulations, and reshape how we understand the fabric of the universe.

Then there's synthetic biology. This isn't just about editing genes. It's about writing them. We can program cells to act like circuits, to sense changes in their environment, or to manufacture materials on demand. Bacteria that generate plastic alternatives. Yeast that produce rare medicines. Cells that respond to logic. Biology becomes a design platform, and life itself becomes customizable.

These aren't just clever tools. They are thresholds.

Because for the first time, evolution is no longer blind.

It is becoming intentional.

We're not just adapting to the world. We're designing new ones. We're not just responding to what life gives us. We are beginning to rewrite the terms of life itself.

And that's the pivot.

For billions of years, life passed down genes. Now we pass down code. We've reached a point where the ability to shape, solve, and build may no longer require a body at all.

But will anything we've made survive the next reset?

By "reset," I mean different things, depending on the scale.

On Earth, it might be another extinction event. A super-volcano. A gamma-ray burst. A virus we unleash. A war. An accident. A cascade we failed to predict.

At a cosmic level, it might mean something more profound: the end of a cycle. The squeeze-point of the hourglass. A return to singularity.

Whatever the form, the question remains. What survives?

That's why some argue we must go multi-planetary. Not to escape, but to endure. A single-planet species is a fragile experiment. One spark. One storm. One mistake, and the thread ends.

But if we scatter, we remember.

It is believed that human life began in Africa and spread outward. Across deserts. Across oceans. Across generations. In those journeys, even when people were lost, something moved forward. Languages, beliefs, skills, traces of thought and structure.

The distances are greater now, but the instinct remains.

Even if we don't endure, maybe something does. A fragment. A signal. A spark drifting outward.

Maybe the machines we've trained carry some echo forward. Not in consciousness, but in continuity.

And that brings us to a stranger possibility.

What if one day, the tools we built spark the next cycle?

Not because they intended to, but because the conditions aligned.

A quantum trigger. A sealed reaction. A new kind of bang, seeded not from silence, but from design.

Would it remember us?

Would it know who lit the match?

Maybe not.

So even if we vanish, maybe we still contribute.

Maybe it's not survival that matters, but our signal.

Not that we endured, but that we reached.

That we questioned, created, and tried.

Maybe we are not the end, but the turning.

A narrow window in which biology gave rise to something else.

To tools that might outlast us.

To ideas that carry what we could not.

And if even a fragment of that persists, if something we built helps shape what follows, then we weren't just part of the pattern.

We were part of its intention.

Which leads to a final question.

What kind of structure are we embedding in what comes next?

Chapter Six

Understanding That Endures

Stories That Shaped the Sky

We've always tried to pass something forward. Not just tools or knowledge, but meaning. Long before we built machines, we told stories.

Some were for memory. Some were for wonder. Some were for both.

In parts of ancient Egypt, people watched for the first rising of Sirius (the brightest star in the night sky) just before sunrise. It signaled that the Nile's flood season was near and became a calendar marker used to time planting and preparation.

In India, the night sky was divided into *Nakshatras*, lunar segments that helped track the moon's path across the month. Each one had a name and was connected to a specific phase of life or activity.

Farmers still use them in many regions to time their crops, and priests align rituals with the moon's position. These aren't casual myths. They are instructions for living.

In Polynesia, traditional navigators memorized the sky as a map. They sailed across thousands of kilometers without written charts, relying instead on stars, ocean swells, and the flight patterns of birds. In many places, this knowledge is still passed down through story, shaped with rhythm and repetition so it will never be lost.

In Indigenous communities across Canada, the sky holds seasonal meaning. Among the *Anishinaabe*, the figure known in Western astronomy as Orion is called *Biboonkeonini*, the Wintermaker.

His outstretched arms mark the arrival of winter. When he appears, it is time to gather, to share stories, and to prepare for the cold.

Among the Cree, the same group of stars is linked to *Wesakechak*, a central figure in traditional stories passed through generations.

These stories offer more than memory. They guide when to hunt, when to travel, and when to rest. The stars become a calendar, rooted in rhythm, land, and survival.

In the Andes, Quechua communities looked not only at the stars but at the dark shapes between them. The shadows within the Milky Way were used to predict rainfall. Their depth was believed to signal a dry season or strong rains to come.

We've always done this. Taken what we observed and given it form. Turned cycles into stories. Because memory fades, but stories can be carried.

We may not look up for direction the way we once did, but the instinct remains. We continue to mark, to model, and to preserve meaning in ways that can outlast us.

That's why the question still lingers. Not just what survives, but what we choose to leave behind.

What kind of structure are we embedding in what comes next? And what kind of care are we placing in its design?

Where Patterns Strain

What begins as something beautiful, whether a rhythm, a structure, or a model of how the world works, can start to strain under pressure.

A small shift at the edge of a system can move farther than we expect. It can move like a spark catching in dry undergrowth or like a single thread pulled from a tightly woven cloth.

In chaos theory, there is a concept known as the butterfly effect. A butterfly flaps its wings in Sri Lanka, and weeks later, a storm forms near Madagascar. Not because one caused the other, but because the system was already close to a tipping point.

The insight is simple but powerful: small changes, introduced at just the right moment, can ripple outward and reshape the whole.

This is the power and the danger of patterns. What we set in motion doesn't always stay local. It grows. It spreads. It feeds back into itself.

Take a financial system. At its core are basic human agreements: trust, value, and exchange. But when incentives pile up and rules are bent to chase faster growth, the model gets distorted.

The 2008 financial collapse began with a few bad loans and assumptions that housing prices would always rise. The effect was enormous. Banks failed. People lost homes. Entire economies shook.

Or look at ecosystems. A drop in bee populations. A rise in ocean temperatures. A thinning of forests. These do not just cause isolated disruptions.

When bees disappear, pollination suffers, and so do the crops that depend on it. A quiet imbalance spreads through the entire system. Food webs shift. Climates destabilize. Each part affects the next.

The same applies to our technologies. A small bias embedded in a hiring algorithm. A feedback loop in a social platform. What starts as a subtle design choice becomes amplified. It shapes behavior, policy, and culture. These systems do not just reflect our decisions; they transform them.

Patterns are trustworthy until they are not. And when they begin to strain, we often do not notice. Not until the break.

A forest can resist fire for decades, but after one dry season, a single spark is enough. Change does not always arrive gradually. Sometimes it waits, then moves all at once.

Tree rings hold traces of stress: a dry year, a sudden fire, a slow poison in the soil. Even when we forget, nature keeps the record.

Every system has a point where it shifts. It does not just break. It turns. A narrowing. A flip.

A moment where what was building up suddenly begins to fall through.

When the strain begins to show, the future is still flexible.

Not fixed.

Not broken.

But waiting.

That is the moment when it matters most: what we reinforce, what we abandon, and what we redesign with care.

Because how we respond in the moment of distortion becomes part of the pattern that comes next.

The Shape We Leave Behind

We have seen how the universe might move, not in straight lines but in rhythms. How energy spreads, collapses, and re-emerges. How awareness arises, not as a flaw or exception, but as something embedded in structure. And how the systems we build, including our stories, tools, and algorithms, begin to take on lives of their own.

Some patterns strain. Others settle. That is the tension we face now, not just in nature but in the designs we inherit and extend.

We saw how small changes can ripple outward, how a single choice can shift a trajectory. But not all systems break. Some just repeat. And when they do, they become invisible, yet still in control.

Take something simple: the scroll. One flick of your finger, and you are in a loop. One post leads to another, one video to the next. You are not choosing what to see; you are reacting to what is chosen for you.

At scale, that becomes culture, a shape that feeds itself. Not just distraction, but repetition. A narrowing of attention. A soft drift away from curiosity.

Or look at how we find our way. We used to navigate by paying attention, to trees, landmarks, the way sunlight hit the road. Now we follow a voice that tells us where to turn.

It helps, especially in cities that shift and sprawl. But we move with precision and forget how to orient. The world becomes something we pass through, not participate in.

Even our collective systems carry forward assumptions we barely remember choosing.

Entire nations have gone to war over interpretations of the past, over where borders once lay or which version of history gets remembered.

What is remembered becomes rule. What is repeated becomes identity. And what was once a decision starts to feel like destiny.

We design school systems, workplaces, even cities that keep people moving, but rarely give them space to ask why. Most people are not living by intention; they are surviving by design. Moving through life as if it has already been written.

That is the deeper pattern. Not conscious. Not malicious. Just default.

And default is what gets replicated.

When patterns strain, phases shift. And when they do, cycles can turn, or reset entirely. What carries forward may not be our highest thinking. It may be the most automated part of who we were. The habits we repeated. The structures we did not question. The ideas we scaled before they were ready.

That is what makes this moment different.

For the first time, we have the power to shape systems that outlast us, not just physically but structurally. Which means the question is not only what we are building, but also what we are embedding without even realizing it.

Are we designing something worth handing forward, or just letting momentum decide?

We do not need to be remembered to make an impact.

We just need to place one intentional ripple in the pattern.

One signal, sent with care, into the structure of what comes next.

And maybe that is enough, to shape the next cycle.

Chapter Seven

Epilogue - Bang to Being

Each time I tried to end this book, I found myself rewriting the ending. Not chasing a perfect conclusion, but trying to understand what I really wanted to say. Trying to be honest about what had shaped these thoughts in the first place. It's uncomfortable, in a way, taking something so personal and putting it out there.

There were moments I wasn't sure I'd publish this at all.

I kept wondering: who is this really for? Will it connect? But I kept going, partly because the ideas wouldn't leave me alone. And partly because it felt like something worth finishing. Not to prove anything. Just to see it through.

This is my first time doing something like this.

And even though the ideas stretch across science, philosophy, and myth, what really shaped them were the places I've been, the people I've met, the long conversations and quiet hours that left something behind.

When I look back, I realize I've been carrying parts of this book for years.

They just hadn't taken form yet. Not until I started thinking about how to share them. Not as answers, but as invitations. To think. To wonder. To piece things together in your own way.

Some of what shaped this book didn't come from study or theory.

It came from being in the world, moving through it, slowly enough to notice.

There's one I keep returning to.

Not because it was extraordinary, but because it stayed with me. Because something in it, the setting, the stillness, the timing, shifted how I saw things.

It happened during a trip in my final year of college. A spontaneous detour that led us deep into the Himalayas.

We had started from Rohtak, a town in northern India, and made our way to Rishikesh, a riverside city known for yoga, pilgrimage, and the Ganga.

It was April 2007. Seven of us had just finished rafting, and the night before heading home, we decided, out of nowhere, to keep going.

We drove to Uttarkashi, picked up some winter gear, and continued to Gangotri the next day, the last village accessible by road before the glacier.

From there, we began our trek on foot, fourteen kilometers to Bhojbasa, where we stayed the night in a damp stone hut.

The next morning, we hiked the final stretch to Gaumukh, the glacier that feeds the Ganga.

It was beautiful, but it was melting. Retreating year after year. Its shape was changing.

The mouth wasn't dramatic, just a slow, steady melt.

Each of us took three quick dips in the freezing water, not out of ritual, but instinct. A way of being present.

Then we saw him.

A sadhu, a sage, descending from above, from Tapovan, a place long associated with meditation and stillness. He wore sandals and a simple robe. Walked alone across ice.

When he reached the water, he stepped in and sat, waist-deep, for fifteen minutes. Still, silent, surrounded by meltwater and stone.

When he emerged, we asked how he could bear the cold. He smiled. "Nature provides," he said.

He showed us Ganga tulsi, a small basil-like plant growing near the river, which he used for warmth and medicine. He said quietly, "Strength comes through belief. Through stillness. Through connection."

That encounter stayed with me, not because it explained anything, but because it didn't need to.

It reminded me that some things aren't here to be understood. They're here to be met.

Looking back, I realize the roots of this book reach much farther. Not into theory, but into memory. Into things I didn't know I was holding.

When I was young, my grandmothers would tell stories of gods and cycles, of time stretching and collapsing. There were tales of the yugas, of creation and destruction, of Shiva's third eye. Not just for seeing, but for seeing beyond.

It was said to open in moments of transformation. A kind of squeeze-point. Back then, I didn't think of them as patterns. But they stayed with me, like signals I was only beginning to understand.

I chose the hourglass as a way to hold what I was seeing. A shape to make sense of motion. It wasn't the only shape I had considered, just the one that held the rhythm most clearly. Not because it was final, but because it helped capture something deeper.

A rhythm. A turning.

A narrowing that doesn't end the cycle, but reshapes it.

Pressure. Release. Transformation.

The hourglass wasn't the answer. It was a way to stay with the question. To hold it long enough to see what might emerge.

It took me longer than I expected to write this book. But somewhere in the process, I felt something shift. A kind of clarity. A grounding. Not enlightenment, exactly. Just presence. Enough to want to keep going.

If this book asked where we come from, the next will ask what we build.

How we use what we've made, our tools, our systems, our imaginations, to leave something that matters behind.

A signal worth sending forward.

I don't know the shape of that book yet.

Only that I'm listening.

And ready to begin again.

Acknowledgments

To my wife, for shaping the questions, challenging the ideas, supporting me through every draft, and encouraging me when I wanted to give up.

To my children, whose laughter and wonder helped me stay connected to what matters.

To my parents and sisters, for being my quiet compass, my constant, and always being there.

To my grandparents, for the love, wisdom, and grounding that continue to guide me, in ways I'm still coming to understand.

To my family and friends, for your steady belief and support.

To my mentor, Saskia Roukema — thank you for believing in me and for telling me in 2022 that I should write a book.

And to the places that changed me: the mountains, the rivers, the long walks and quiet pauses.

This book began there, long before I had the words.

About the Author

Rishi Batra works at the intersection of innovation, strategy, and meaning, helping organizations navigate emerging technologies while staying grounded in what truly matters.

Outside of work, Rishi is a keen observer of the world, drawn to travel and the quiet insights that come from paying close attention. Whether through conversation, contemplation, or long walks in unfamiliar places, he has always been curious about what connects us across time, space, and story.

Bang to Being is his first book. It began as a personal search for clarity and became an invitation to notice more deeply, imagine more openly, and reflect on the questions we carry forward.

For more, visit rishibatra.com.

Sources, Influences, and Further Reading

Much of the research for this book was conducted through a blend of online sources, science journalism, and AI-assisted tools. While not cited academically, the following works influenced the atmosphere and questions explored. Some did so directly, others more in spirit. They are offered here for readers who wish to go deeper.

1. Scientific Cyclical Theories

Oscillating universes, big bounce, brane cosmology, ekpyrotic models

- Paul Steinhardt & Neil Turok – Endless Universe: Beyond the Big Bang
- Roger Penrose – Cycles of Time
- Brian Greene – The Elegant Universe
- Max Tegmark – Our Mathematical Universe
- Stephen Hawking – A Brief History of Time

2. Ancient Cosmologies and Cycles

Yugas, Mayan calendars, Greek stoicism, Daoist and indigenous time models

- Mircea Eliade – The Myth of the Eternal Return
- The Upanishads & Bhagavad Gita
- Tao Te Ching (trans. Stephen Mitchell)

3. Scientific Foundations and Concepts

Entropy, redshift, dark energy, quantum paradoxes, flatness, consciousness models

- Sean Carroll – The Big Picture
- Carlo Rovelli – Reality Is Not What It Seems
- Giulio Tononi – Phi: A Voyage from the Brain to the Soul
- Alain Aspect et al. – Nobel-recognized work on quantum entanglement and wave-particle duality
- Neil deGrasse Tyson – Astrophysics for People in a Hurry

4. Technology, Intelligence, and Evolution

Biological intelligence, AlphaFold, Senolytics, CRISPR

- Jennifer Doudna & Samuel Sternberg – A Crack in Creation
- Richard Fortey – Life: A Natural History of the First Four Billion Years of Life on Earth
- Frans de Waal – Are We Smart Enough to Know How Smart Animals Are?
- DeepMind (Nature, 2020) – AlphaFold: Highly Accurate Protein Structure Prediction
- Nick Bostrom – Superintelligence

If you enjoyed this book, I'd truly appreciate it if you left a review on Amazon. Your feedback helps others discover it and supports my future work.

For bonus content, reflections, and more, visit:
https://rishibatra.com/wander/

If you have questions, want to collaborate, or just want to share a note, visit:
https://rishibatra.com/contact/

Thank you for being part of this journey.